# Algebra: A Complete Course

## Module A – Course Notes

VideoText *Interactive*

Send all inquiries to:
VideoText*Interactive*
P.O. Box 19761
Indianapolis, IN 46219



ISBN 1-887891-37-4
9 10 11 12 - RPInc - 10 09 08 07

# Table of Contents

## Unit I – The Structure of Mathematics

COURSE NOTE 1

# 5 PARTS OF MATHEMATICAL SPEECH

1. Number Symbols "things" (nouns) — $\{1, 2, 3, \frac{4}{5}, .9, \ldots\}$

2. Operation Symbols "actions" (verbs) — $\{+, -, \times, \div, \ldots\}$

3. Relation Symbols "comparisons" — $\{=, >, <, \ldots\}$

4. Grouping Symbols "associations" (punctuation) — $\{(\ ), [\ ], \{\ \}, \ldots\}$

5. Placeholder Symbols "hold the place" (pronouns) — $\{a, b, c, ?, \square, \ldots\}$

COURSE NOTE 2

# TYPES OF MATHEMATICAL EXPRESSIONS

## 1. Closed Phrase

*Ex.* $7 + 9$

no relation
no placeholder

So, Evaluate

*Ex.* 16

## 2. Open Phrase

*Ex.* $7 + n$

no relation
has placeholder

So, Substitute and Evaluate

*Ex.* Domain $= \{0, 1, 2\}$ → Range $= \{7, 8, 9\}$

## 3. Closed Sentence

*Ex.* $7 + 9 = 17$

has relation
no placeholder

So, True or False

*Ex.* False

## 4. Open Sentence

*Ex.* $7 + n = 17$

has relation
has placeholder

So, Substitute and True or False

*Ex.* Replacement $= \{4, 5, 6\}$ → Solution $= \{\ \}$

COURSE NOTE 3

# TRANSLATION OF MATH SYMBOLS

1. Number Symbols $\{1, 2, 3, \frac{4}{5}, .9, \ldots\}$

   $3 \rightarrow$ three

2. Operation Symbols $\{+, -, \times, \div, \ldots\}$

| | | | |
|---|---|---|---|
| + | add | rise | increase |
| | sum | climb | combine |
| | plus | total | number more than |
| | gain | | |
| – | subtract | loss | take away |
| | difference | fall | deduct |
| | minus | decrease | number less than |
| x or · | times | of | twice |
| | multiply | double | |
| | product | triple | |
| ÷ | divide | goes into | |
| | quotient | fraction | |

3. Relation Symbols $\{=, >, <, \ldots\}$

   $=$ is equal to, is the same as
   $>$ is greater than, is more than
   $<$ is less than, is smaller than

COURSE NOTE 4

# TRANSLATION OF MATH SYMBOLS

4. Grouping Symbols $\{(\ ), [\ ], \{\ \}, \ldots\}$

( ) parentheses
[ ] brackets
{ } braces

| | |
|---|---|
| the quantity | "product" |
| "sum" | "quotient" |
| "difference" | |

5. Placeholder Symbols $\{a, b, c, \ldots, ?, \ldots, \square, \ldots\}$

| | |
|---|---|
| a "number" | the "distance" |
| the "unknown" | the "weight" |
| the "age" | ⋮ |

COURSE NOTE 5

# TRANSLATION OF MATH SYMBOLS

The sum of a number and 7 | is equal to | 20.

$$n + 7 \quad = \quad 20$$

4 | is | 1 less than | the quotient of a number and 9.

$$4 = \frac{n}{9} - 1$$

39 | is greater than | the result of multiplying | the sum of 6 and some number | by 3.

$$39 > (6 + n) \cdot 3$$

COURSE NOTE 6

# WELL-DEFINED OPERATIONS

1. **Existence**

2. **Uniqueness**

3. **Closure**

COURSE NOTE 7

# DEVELOPMENT OF THE NUMBER SYSTEM

Natural Numbers $\{1, 2, 3, \ldots\}$

Whole Numbers $\{0, 1, 2, 3, \ldots\}$

Integers $\{\ldots, {}^{-}3, {}^{-}2, {}^{-}1, 0, 1, 2, 3, \ldots\}$

Rational Numbers $\left\{\frac{a}{b} \mid a, b \text{ are integers}\right\}$

Real Numbers $\{\text{rationals}\} \cup \{\text{irrationals}\}$

All Numbers $\{\text{reals}\} \cup \{\text{complex}\}$

COURSE NOTE 8

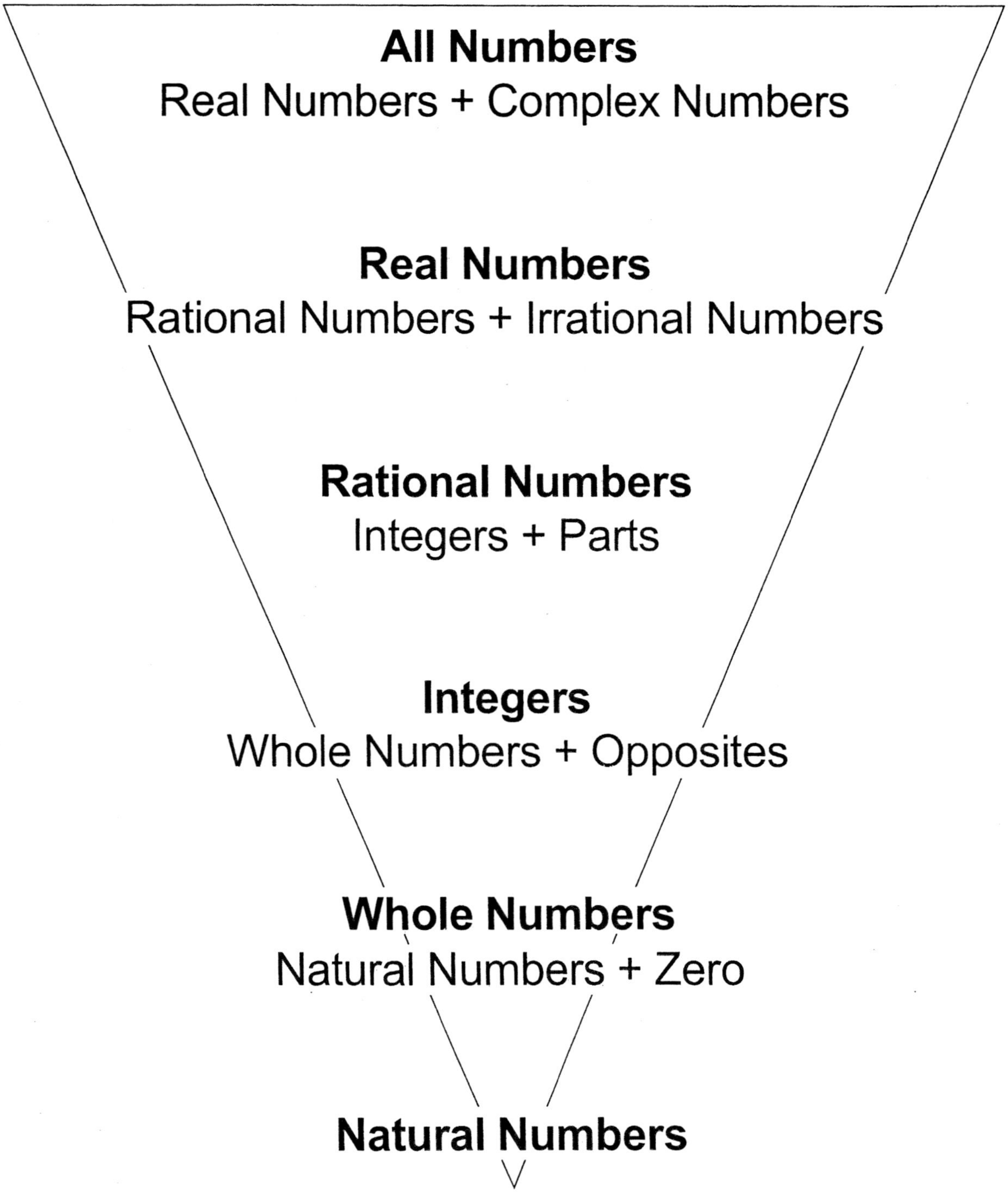

COURSE NOTE 9

# FRACTIONS

## "PART OF SOMETHING"
## FRACTURE

| | | | |
|---|---|---|---|
| $a$ | – **numerator** → | numeral<br>numerous<br>enumerate | → "How many?" |
| $b$ | – **denominator** → | nominate<br>denomination<br>nominal | → "What kind?" |

COURSE NOTE 10

# EQUI VALENT FORMS

EQUI → equal
VALENT → value

$$\frac{24}{30}$$

$$\frac{2 \cdot 3 \cdot 2 \cdot 2}{2 \cdot 3 \cdot 5}$$

$$\frac{2}{2} \cdot \frac{3}{3} \cdot \frac{2 \cdot 2}{5}$$

$$1 \cdot 1 \cdot \frac{4}{5}$$

$$\frac{4}{5}$$

COURSE NOTE 11

# FRACTIONS TO DECIMALS

$$\frac{139}{4} \rightarrow 139 \div 4 \rightarrow 4\overline{)139.00}$$

$$\begin{array}{r} 34.75 \\ 4\overline{)139.00} \\ \underline{12}\phantom{.00} \\ 19\phantom{.00} \\ \underline{16}\phantom{.00} \\ 3\,0\phantom{0} \\ \underline{2\,8}\phantom{0} \\ 20 \\ \underline{20} \end{array}$$

**TERMINATING**

$$\frac{149}{11} \rightarrow 149 \div 11 \rightarrow 11\overline{)149.0000}$$

$$\begin{array}{r} 13.5454 \\ 11\overline{)149.0000} \\ \underline{11}\phantom{9.0000} \\ 39\phantom{.0000} \\ \underline{33}\phantom{.0000} \\ 6\,0\phantom{000} \\ \underline{5\,5}\phantom{000} \\ 50\phantom{00} \\ \underline{44}\phantom{00} \\ 60\phantom{0} \\ \underline{55}\phantom{0} \\ 50 \\ \underline{44} \\ \vdots \end{array} = 13.\overline{54}$$

**REPEATING**

COURSE NOTE 12

# DECIMALS TO FRACTIONS

## TERMINATING

*Ex.* 6.784 (read it)

$$6\frac{784}{1000} \text{ or } \frac{6784}{1000}$$

---

## REPEATING

*Ex.* $4.626262... = 4.\overline{62}$ (eliminate repetend)

$$\begin{aligned} 100f &= 462.626262... \\ {}^{-}1f &= -4.626262... \\ \hline 99f &= 458. \\ f &= \frac{458}{99} \rightarrow 4.\overline{62} \end{aligned}$$

*Ex.* $3.7514514514 = 3.7\overline{514}$

$$\begin{aligned} 1000f &= 3751.451451... \\ {}^{-}1f &= 3.751451... \\ \hline 999f &= 3747.7 \\ f &= \frac{3747.7}{999} \times \frac{10}{10} = \frac{37477}{9990} \rightarrow 3.7\overline{514} \end{aligned}$$

COURSE NOTE 13

# PERCENT

FRACTION

$$\frac{48}{100}$$

DECIMAL

.48

PERCENT

48%

---

FRACTION

$$\frac{56}{1000} \cdot \frac{\frac{1}{10}}{\frac{1}{10}} \rightarrow \frac{5.6}{100}$$

DECIMAL

.056

PERCENT

5.6%

---

FRACTION

$$\frac{37.5}{100} \cdot \frac{10}{10} = \frac{375}{1{,}000}$$

DECIMAL

.375

PERCENT

37.5%

COURSE NOTE 14

# PRIMES, COMPOSITES, & FACTORING

**PRIME NUMBERS**

[ Primary grades
Prime time
Primer coat ]

"First"
"Beginning"

**COMPOSITE NUMBERS**

[ Compose
Composition ]

"Several parts"

**FACTORING**

[ Factory
Faction ]

"Parts"

Ex. 24

| Factored Forms | | | |
|---|---|---|---|
| | $2 \cdot 12$ | $3 \cdot 8$ | $4 \cdot 6$ |
| | $2 \cdot 2 \cdot 6$ | $3 \cdot 2 \cdot 4$ | $2 \cdot 2 \cdot 2 \cdot 3$ |
| | $2 \cdot 2 \cdot 2 \cdot 3$ | $3 \cdot 2 \cdot 2 \cdot 2$ | |

Prime Factored Form **$2 \cdot 2 \cdot 2 \cdot 3$**

COURSE NOTE 15

# ERATOSTHENES' SIEVE

| | | | | | | | | | |
|---|---|---|---|---|---|---|---|---|---|
| 1 | 2 | 3 | 4 | 5 | 6 | 7 | 8 | 9 | 10 |
| 11 | 12 | 13 | 14 | 15 | 16 | 17 | 18 | 19 | 20 |
| 21 | 22 | 23 | 24 | 25 | 26 | 27 | 28 | 29 | 30 |
| 31 | 32 | 33 | 34 | 35 | 36 | 37 | 38 | 39 | 40 |
| 41 | 42 | 43 | 44 | 45 | 46 | 47 | 48 | 49 | 50 |
| 51 | 52 | 53 | 54 | 55 | 56 | 57 | 58 | 59 | 60 |
| 61 | 62 | 63 | 64 | 65 | 66 | 67 | 68 | 69 | 70 |
| 71 | 72 | 73 | 74 | 75 | 76 | 77 | 78 | 79 | 80 |
| 81 | 82 | 83 | 84 | 85 | 86 | 87 | 88 | 89 | 90 |
| 91 | 92 | 93 | 94 | 95 | 96 | 97 | 98 | 99 | 100 |

⋮

COURSE NOTE 16

# LEAST COMMON MULTIPLE

For 8 and 12 (by listing)

Multiples of 8 – 8, 16, 24, 32, 40, 48, 56, 64, 72, 80, 88, 96, ...
Multiples of 12 – 12, 24, 36, 48, 60, 72, 84, 96, 108, ...

Common Multiples of 8 and 12 – 24, 48, 72, 96, ...

LEAST COMMON MULTIPLE OF 8 and 12 is (24)

---

For 8 and 12 (by factoring)

**8** — $2 \cdot 2 \cdot 2$

**12** — $2 \cdot 2 \cdot 3$

**LCM** $2 \cdot 2 \cdot 2 \cdot 3 = \mathbf{24}$

---

For 48 and 60 (by factoring)

**48** — $2 \cdot 2 \cdot 2 \cdot 2 \cdot 3$

**60** — $2 \cdot 2 \cdot 3 \cdot 5$

**LCM** $2 \cdot 2 \cdot 2 \cdot 2 \cdot 3 \cdot 5 = \mathbf{240}$

COURSE NOTE 17

# GREATEST COMMON FACTOR

For 60 and 72 (by listing)

Factors of 60 – 1, 2, 3, 4, 5, 6, 10, 12, 15, 20, 30, 60
Factors of 72 – 1, 2, 3, 4, 6, 8, 9, 12, 18, 24, 36, 72

Common Factors of 60 and 72 – 1, 2, 3, 4, 6, 12

GREATEST COMMON FACTOR OF 60 and 72 is (12)

---

For 60 and 72 (by factoring)

**60**

$\underline{2 \cdot 2 \cdot 3} \cdot 5$

**72**

$2 \cdot \underline{2 \cdot 2 \cdot 3} \cdot 3$

**GCF** $\underline{2 \cdot 2 \cdot 3}$ = (**12**)

---

For 252 and 294 (by factoring)

**252**

$2 \cdot \underline{2 \cdot 3} \cdot 3 \cdot \underline{7}$

**294**

$\underline{2 \cdot 3 \cdot 7} \cdot 7$

**GCF** $\underline{2 \cdot 3 \cdot 7}$ = (**42**)

# ORDER OF OPERATIONS

1. Grouping Symbols
2. Multiplication and Division
3. Addition and Subtraction

---

$$2 + 3 \cdot 4$$
$$2 + 12$$
$$14$$

$$12 \cdot 6 \div 3 \cdot 2 \div 48$$
$$72 \div 3 \cdot 2 \div 48$$
$$24 \cdot 2 \div 48$$
$$48 \div 48$$
$$1$$

---

$$11 + 2 \cdot (6+4) - 3 \cdot (5 \cdot 2 - 6)$$
$$11 + 2 \cdot 10 - 3 \cdot (10 - 6)$$
$$11 + 2 \cdot 10 - 3 \cdot 4$$
$$11 + 20 - 12$$
$$31 - 12$$
$$19$$

---

$$\frac{2 \cdot (5 + 6)}{15 \div 3 + 2 \cdot 3} = \frac{2 \cdot 11}{5 + 6} = \frac{22}{11} = 2$$

COURSE NOTE 19

# PROPERTIES OF OPERATIONS

## COMMUTATIVE ("moving" numbers)

- of addition $a + b = b + a$
- of multiplication $a \cdot b = b \cdot a$

## ASSOCIATIVE ("grouping" numbers)

- of addition $(a + b) + c = a + (b + c)$
- of multiplication $(a \cdot b) \cdot c = a \cdot (b \cdot c)$

## DISTRIBUTIVE ("handing out" numbers)

- of mult. to addition $a \cdot (b + c) = (a \cdot b) + (a \cdot c)$
- of mult. to subtraction $a \cdot (b - c) = (a \cdot b) - (a \cdot c)$

COURSE NOTE 20

# PROPERTIES OF OPERATIONS FOR 0

Identity Property of Addition

"Any number + 0 is the same number"

(0 is the identity element for addition)

Identity Property of Subtraction

"Any number – 0 is the same number"

(0 is the identity element for subtraction)

Multiplication Property of 0

"Any number · 0 is 0"

Division by 0 is meaningless and not allowed

Additive Inverse Property

"Any number + its additive inverse (opposite) is 0"

(0 is the inverse element for addition)

COURSE NOTE 21

# PROPERTIES OF OPERATIONS FOR 1

<u>Identity Property of Multiplication</u>

"Any number $\cdot$ 1 is the same number"

(1 is the identity element for multiplication)

<u>Identity Property of Division</u>

"Any number $\div$ 1 is the same number"

(1 is the identity element for division)

<u>Multiplicative Inverse Property</u>

"Any number $\cdot$ its multiplicative inverse (reciprocal) is 1"

(1 is the inverse element for multiplication)

COURSE NOTE 22

# MULTIPLICATION OF FRACTIONS

$$\frac{a}{b} \cdot \frac{c}{d} = \frac{a \cdot c}{b \cdot d}$$

$$\frac{3}{4} \cdot \frac{8}{21}$$

$$\frac{3 \cdot 8}{4 \cdot 21}$$

$$\frac{24}{84}$$

$$\frac{\cancel{3} \cdot \cancel{2} \cdot \cancel{2} \cdot 2}{\cancel{2} \cdot \cancel{2} \cdot \cancel{3} \cdot 7}$$

$$\frac{2}{7}$$

$$\frac{3}{4} \cdot \frac{8}{21}$$

$$\frac{3 \cdot 8}{4 \cdot 21}$$

$$\downarrow$$

$$\frac{\cancel{3} \cdot \cancel{2} \cdot \cancel{2} \cdot 2}{\cancel{2} \cdot \cancel{2} \cdot \cancel{3} \cdot 7}$$

$$\frac{2}{7}$$

COURSE NOTE 23

# ADDITION OF FRACTIONS

$$\frac{a}{b}+\frac{c}{b}=\frac{a+c}{b}$$

"SAME KIND" $\frac{2}{7}+\frac{4}{7}=\frac{6}{7}$

---

"DIFFERENT KIND"

| | | | |
|---|---|---|---|
| | $\frac{1}{3} + \frac{1}{4}$ | | $\frac{3}{8} + \frac{5}{12}$ |
| L.C.M. | (3) (2)(2) ↘ ↙ $3\cdot 2\cdot 2$ 12 | L.C.M. | (2)(2)(2) (2)(2)(3) ↘ ↙ $2\cdot 2\cdot 2\cdot 3$ 24 |
| WHICH? | $\frac{1}{3}\cdot 1+\frac{1}{4}\cdot 1$ | WHICH? | $\frac{3}{8}\cdot 1+\frac{5}{12}\cdot 1$ |
| HOW? | $\frac{1}{3}\cdot\frac{4}{4}+\frac{1}{4}\cdot\frac{3}{3}$ | HOW? | $\frac{3}{8}\cdot\frac{3}{3}+\frac{5}{12}\cdot\frac{2}{2}$ |
| CHANGE | $\frac{4}{12}+\frac{3}{12}$ | CHANGE | $\frac{9}{24}+\frac{10}{24}$ |
| ADD | $\frac{7}{12}$ | ADD | $\frac{19}{24}$ |

COURSE NOTE 24

# SUBTRACTION OF FRACTIONS

$$\frac{a}{b} - \frac{c}{b} = \frac{a-c}{b}$$

$$\frac{5}{6} \quad - \quad \frac{3}{18}$$

$$(2)(3) \quad (2)(3)(3)$$

L.C.M.

$$2 \cdot 3 \cdot 3$$

$$18$$

WHICH? $\frac{5}{6} \cdot 1 - \frac{3}{18}$

HOW? $\frac{5}{6} \cdot \frac{3}{3} - \frac{3}{18}$

CHANGE $\frac{15}{18} - \frac{3}{18}$

SUBTRACT $\frac{12}{18} = \frac{\cancel{2}^{1} \cdot 2 \cdot \cancel{3}^{1}}{\cancel{2}_{1} \cdot 3 \cdot \cancel{3}_{1}} = \frac{2}{3}$

COURSE NOTE 25

# DIVISION OF FRACTIONS

$$\frac{a}{b} \div \frac{c}{d} = \frac{\frac{a}{b}}{\frac{c}{d}} \left[\frac{\cdot\frac{d}{c}}{\cdot\frac{d}{c}}\right] = \frac{\frac{a}{b}\cdot\frac{d}{c}}{1} = \frac{a\cdot d}{b\cdot c}$$

---

$$\frac{5}{6} \div \frac{3}{8} = \frac{\frac{5}{6}}{\frac{3}{8}} \left[\frac{\cdot\frac{8}{3}}{\cdot\frac{8}{3}}\right] = \frac{\frac{5}{6}\cdot\frac{8}{3}}{\frac{3}{8}\cdot\frac{8}{3}} = \frac{\frac{40}{18}}{1}$$

$$= \frac{\cancel{2}\cdot 2\cdot 2\cdot 5}{\cancel{2}\cdot 3\cdot 3}$$

$$= \frac{20}{9} \text{ or } 2\frac{2}{9}$$

COURSE NOTE 26

# ADDING AND SUBTRACTING DECIMALS

$$.4 + .5 \quad = \quad \begin{array}{r} .4 \\ +\ .5 \\ \hline .9 \end{array}$$

$$.3 - .21 \quad = \quad \begin{array}{r} .30 \\ -\ .21 \\ \hline .09 \end{array}$$

# MULTIPLYING DECIMALS

$$.6 \times .7 \quad = \quad \begin{array}{rl} .6 & \frac{6}{10} \\ \times\ \ .7 & \frac{7}{10} \\ \hline .42 & \frac{42}{100} \end{array}$$

$$.4 \times .37 \quad = \quad \begin{array}{rl} .4 & \frac{4}{10} \\ \times\ \ .37 & \frac{37}{100} \\ \hline .148 & \frac{148}{1000} \end{array}$$

$$1.04 \times .67 \quad = \quad \begin{array}{rl} 1.04 & \frac{104}{100} \\ \times\ \ .62 & \frac{62}{100} \\ \hline 208 & \\ 624\ \ & \\ \hline .6448 & \frac{6448}{10000} \end{array}$$

COURSE NOTE 27

# DIVIDING DECIMALS

$5.67 \div 7$

$$\begin{array}{r} .81 \\ 7\overline{)5.67} \\ \underline{5\,6\phantom{7}} \\ 7 \\ \underline{7} \end{array}$$

$43.74 \div .3$

$$\frac{43.74}{.3} \times \frac{10}{10} = \frac{437.4}{3}$$

$$.3\overline{)43.74}$$

$$\begin{array}{r} 145.8 \\ 3\overline{)437.4} \\ \underline{3\phantom{37.4}} \\ 13\phantom{7.4} \\ \underline{12\phantom{7.4}} \\ 17\phantom{.4} \\ \underline{15\phantom{.4}} \\ 24 \\ \underline{24} \end{array}$$

COURSE NOTE 28

# VECTOR

1. Direction (Right or Left) $\rightarrow$ Sign (Positive or Negative)

2. Length (Number of Units) $\rightarrow$ Numerical Value (Absolute Value)

$$\left|{}^{+}8\right| = 8$$

$$\left|{}^{-}8\right| = 8$$

$$\left|{}^{+}6\right| = \left|{}^{-}6\right| = 6$$

COURSE NOTE 29

# ADDING VECTORS

$$(^{+}3)+(^{+}6)+(^{-}2)+(^{-}5)={}^{+}2$$

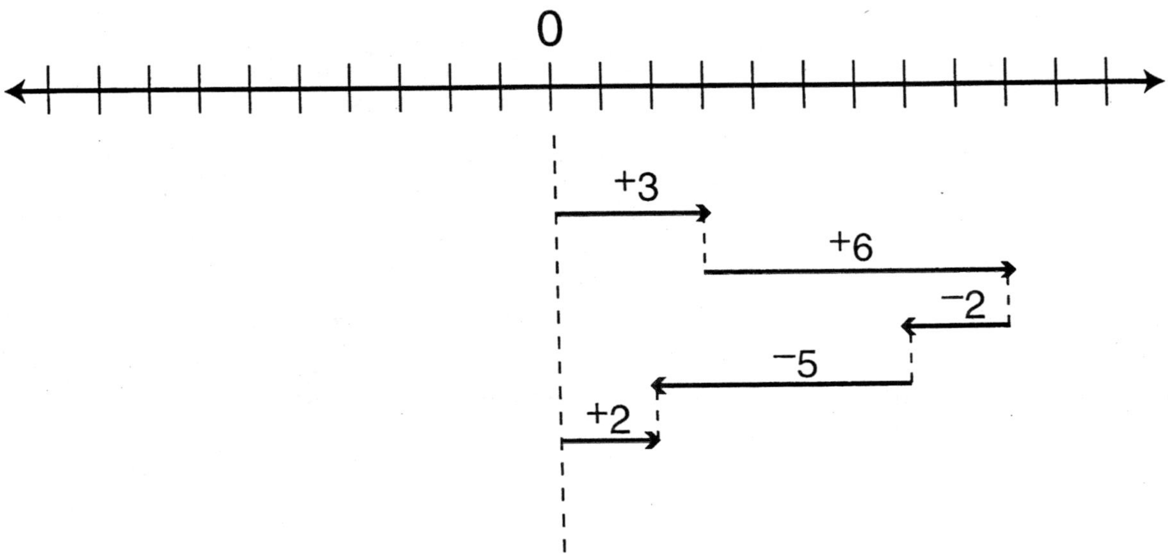

COURSE NOTE 30

# ADDITION OF SIGNED NUMBERS

$(+)+(+)=$ → sign $(+)$; → absolute value (sum of absolute values)

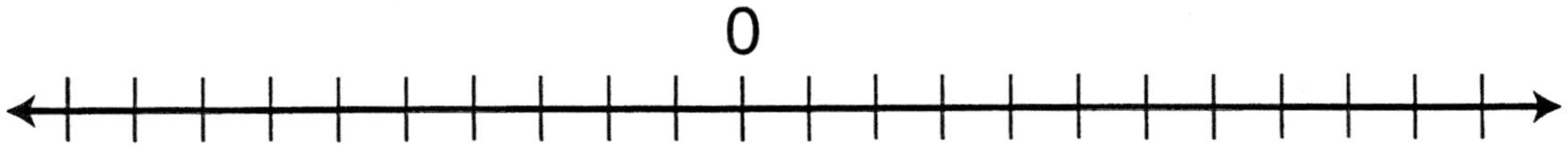


$(^{+}3)+(^{+}4)={}^{+}7$

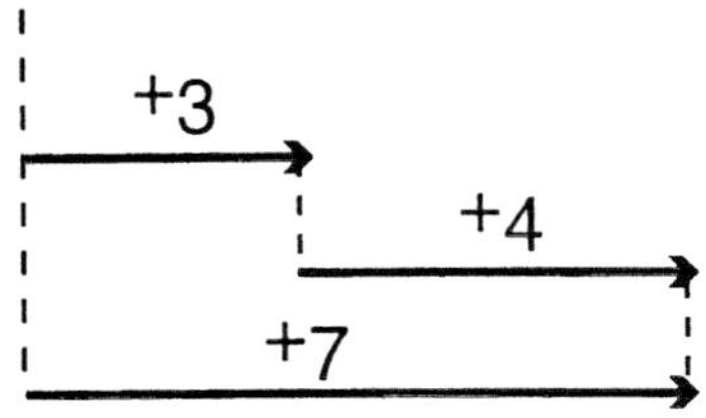


$(-)+(-)=$ → sign $(-)$; → absolute value (sum of absolute values)

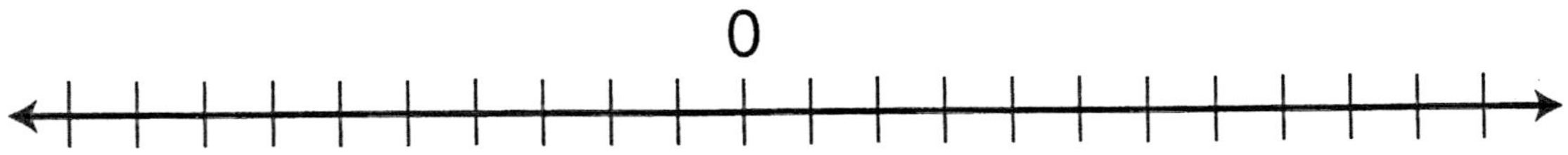


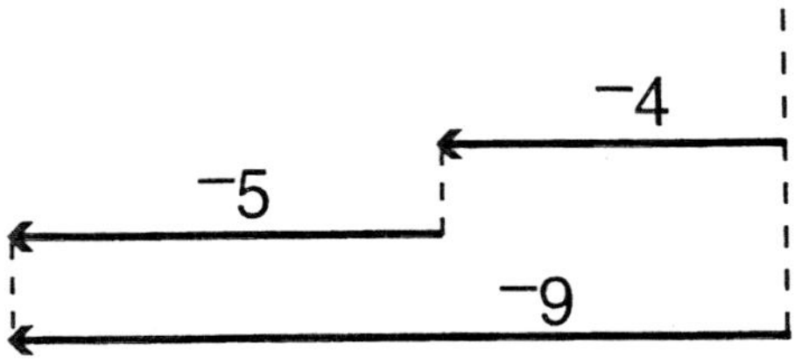


$(^{-}4)+(^{-}5)={}^{-}9$

# ADDITION OF SIGNED NUMBERS

$$\begin{bmatrix}(+)+(-)\\(-)+(+)\end{bmatrix} =$$

sign (+) or (−) (greater absolute value)

absolute value (difference of absolute values)

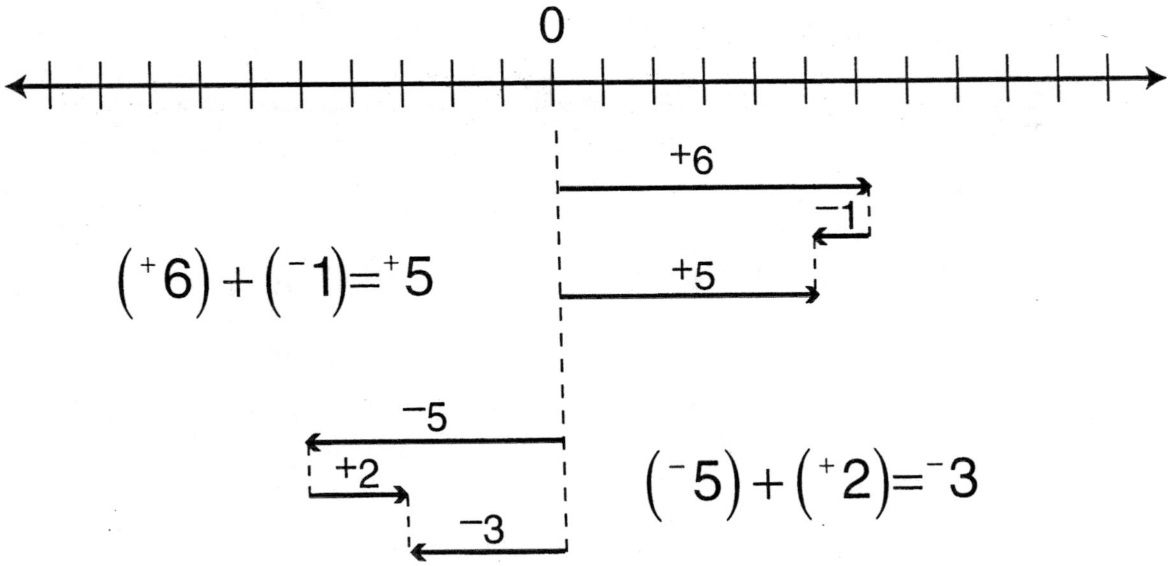


$$({}^{+}6)+({}^{-}1)={}^{+}5$$

$$({}^{-}5)+({}^{+}2)={}^{-}3$$

COURSE NOTE 32

# SUBTRACTION OF SIGNED NUMBERS

## Uses of the "DASH"

| | | |
|---|---|---|
| $5 - 2$ | $\rightarrow$ | Subtract |
| $^{-}4$ | $\rightarrow$ | Negative |
| $-3$ | $\rightarrow$ | "The Opposite Of" |

$$-\left(-^{+}2\right) = ^{+}2$$

$$-\left(-\left[-\left(-\left[-^{-}4\right]\right)\right]\right) = ^{+}4$$

**"Every two dashes cancel each other out"**

---

$(^{+}7) - (^{+}3)$

$^{+}7 + (^{-}3) \quad = \quad ^{+}4$

$(^{-}6) - (^{+}2)$

$^{-}6 + ^{-}2 \quad = \quad ^{-}8$

$(^{+}9) - (^{-}4)$

$^{+}9 + ^{+}4 \quad = \quad ^{+}13$

$(^{-}3) - (^{-}8)$

$^{-}3 + ^{+}8 \quad = \quad ^{+}5$

"Just find the **opposite** and **add**"

$-\left(^{-}3\right) - \left(-\left[^{-}4\right]\right)$

$\left(^{+}3\right) + \left(^{-}4\right)$

$^{-}1$

COURSE NOTE 33

# MULTIPLICATION OF SIGNED NUMBERS

$(+)\cdot(+) =$ sign $(+)$; absolute value (product of absolute values)

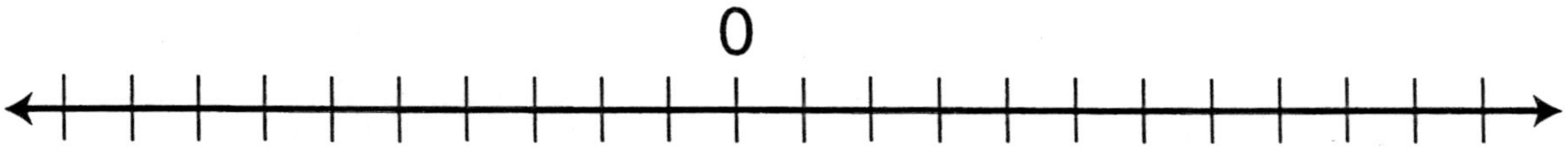


$({}^{+}3)\cdot({}^{+}4)={}^{+}12$

$({}^{+}4)+({}^{+}4)+({}^{+}4)$

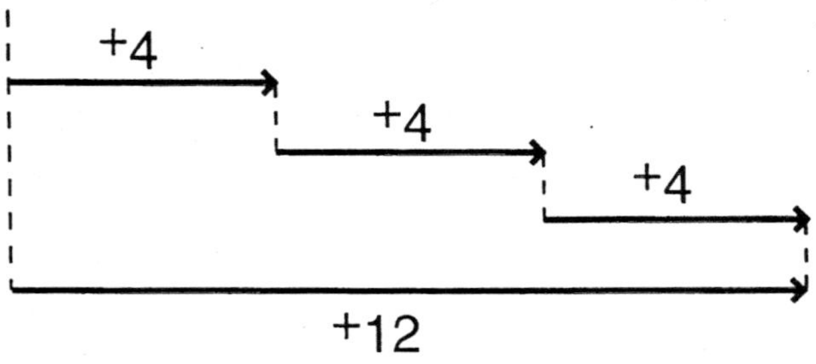


---

$\left|\begin{matrix}(+)\cdot(-)\\(-)\cdot(+)\end{matrix}\right| =$ sign $(-)$; absolute value (product of absolute values)

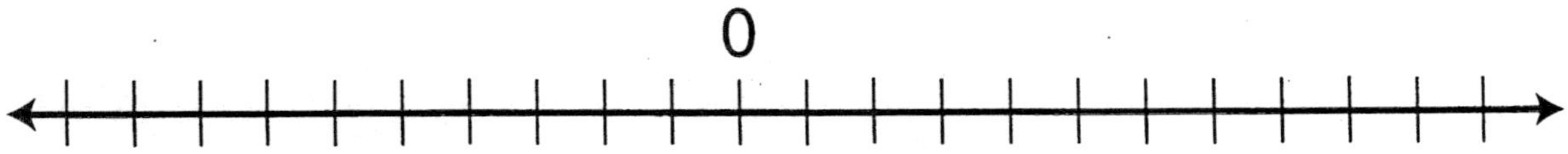


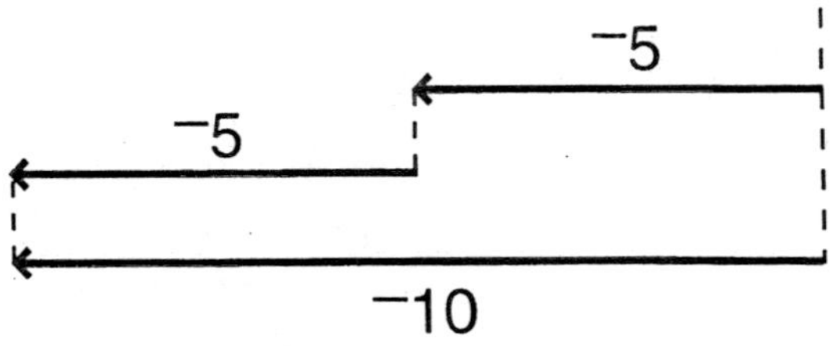


$({}^{+}2)\cdot({}^{-}5)={}^{-}10$

$({}^{-}5)+({}^{-}5)$

COURSE NOTE 34

# MULTIPLICATION OF SIGNED NUMBERS

$(-)\cdot(-) =$ sign $(+)$ ; absolute value $\left(\text{product of absolute values}\right)$

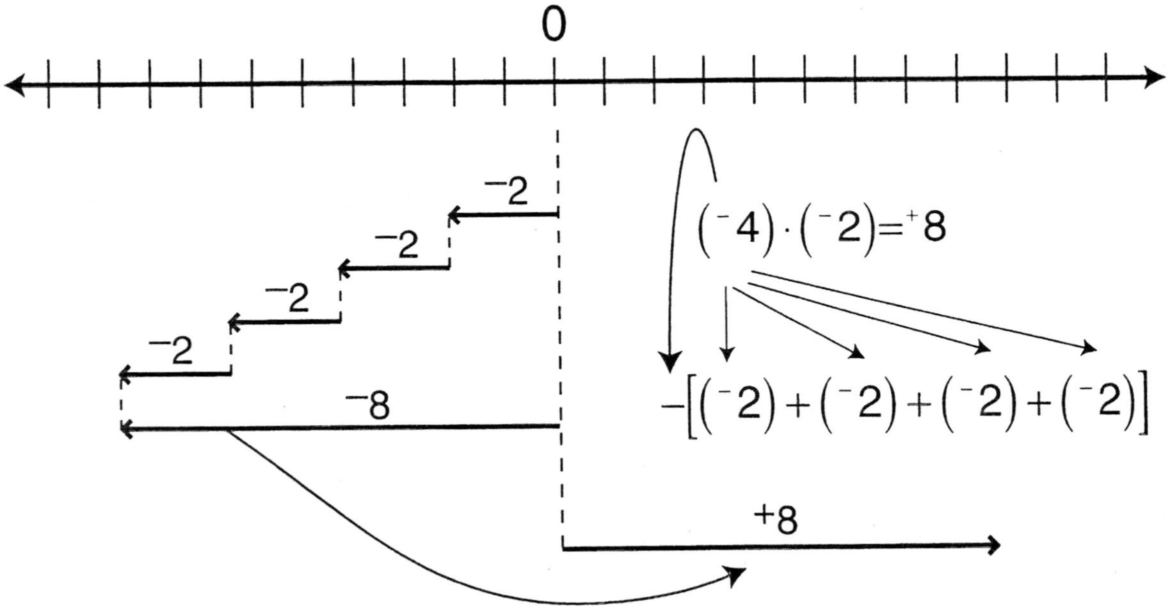

COURSE NOTE 35

# DIVISION OF SIGNED NUMBERS

$(+) \div (+) = $ sign $(+)$, absolute value (quotient of absolute values)

$$\left({}^{+}12\right) \div \left({}^{+}3\right)$$

$$\frac{{}^{+}12}{1} \div \frac{{}^{+}3}{1}$$

$$\frac{{}^{+}12}{1} \cdot \frac{{}^{+}1}{3} = \frac{{}^{+}12}{3} = {}^{+}4$$

---

$\begin{bmatrix} (+) \div (-) \\ (-) \div (+) \end{bmatrix} = $ sign $(-)$, absolute value (quotient of absolute values)

$$\left({}^{+}15\right) \div \left({}^{-}5\right)$$

$$\frac{{}^{+}15}{1} \div \frac{{}^{-}5}{1}$$

$$\frac{{}^{+}15}{1} \cdot \frac{-1}{5} = \frac{{}^{-}15}{5} = {}^{-}3$$

---

$(-) \div (-) = $ sign $(+)$, absolute value (quotient of absolute values)

$$\left({}^{-}10\right) \div \left({}^{-}5\right)$$

$$\frac{{}^{-}10}{1} \div \frac{{}^{-}5}{1}$$

$$\frac{{}^{-}10}{1} \cdot \frac{-1}{5} = \frac{{}^{+}10}{5} = {}^{+}2$$

COURSE NOTE 36

# NUMBER LINE

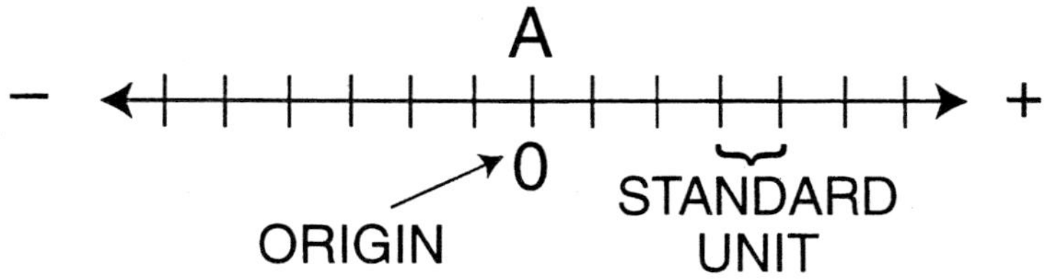


0 is the coordinate of A

A is the graph of 0

| | |
|---|---|
| Natural Numbers | $\{1, 2, 3, \ldots\}$ |
| Whole Numbers | $\{0, 1, 2, 3, \ldots\}$ |
| Integers | $\{\ldots, {}^{-}3, {}^{-}2, {}^{-}1, 0, 1, 2, 3, \ldots\}$ |
| Rational Numbers | $\left\{\frac{a}{b}\right\}$ |

## Order on the Number Line

| $<$ | $=$ | $>$ |
|---|---|---|
| is less than | is equal to | is greater than |
| $2 < 5$ | $\frac{12-6}{2} = \frac{7+5}{4}$ | ${}^{-}1 > {}^{-}3$ |
| | $3 = 3$ | |

COURSE NOTE 37

# DEFINITION OF EQUALITY

1. $n = n$ (Reflexive)
2. if $x = y$, then $y = x$ (Symmetric)
3. if $a = b$, and $b = c$, then $a = c$ (Transitive)

## Properties of Equality

| | | | |
|---|---|---|---|
| Given | $3+1$ | $=$ | $4$ |
| Add 5 | $3+1+5$ | $=$ | $4+5$ |
| Subtract 1 | $3+1-1$ | $=$ | $4-1$ |
| Multiply 5 | $5(3+1)$ | $=$ | $5(4)$ |
| Divide 2 | $\frac{3+1}{2}$ | $=$ | $\frac{4}{2}$ |

"You may do **anything** to an equation (except divide by 0), as long as you do it to **both** sides."

COURSE NOTE 38

# PROPERTIES OF INEQUALITY

| | | | |
|---|---|---|---|
| GIVEN | $2+3$ | $<$ | $6$ |
| ADD 4 | $2+3+4$ | $<$ | $6+4$ |

| | | | |
|---|---|---|---|
| GIVEN | $6+2$ | $>$ | $7$ |
| SUBTRACT 2 | $6+2-2$ | $>$ | $7-2$ |

| | | | |
|---|---|---|---|
| GIVEN | $4$ | $<$ | $5$ |
| MULTIPLY 3 | $3\,(4)$ | $<$ | $3\,(5)$ |

| | | | |
|---|---|---|---|
| GIVEN | $8$ | $>$ | $6$ |
| DIVIDE 2 | $\frac{8}{2}$ | $>$ | $\frac{6}{2}$ |

| | | | |
|---|---|---|---|
| GIVEN | $4$ | $<$ | $6$ |
| MULTIPLY $^{-}3$ | $^{-}3\,(4)$ | $>$ | $^{-}3(6)$ |
| | $^{-}12$ | $>$ | $^{-}18$ |
| DIVIDE $^{-}2$ | $\frac{^{-}12}{^{-}2}$ | $<$ | $\frac{^{-}18}{^{-}2}$ |
| | $6$ | $<$ | $9$ |

COURSE NOTE 39

"You may do **anything** to an equation or inequality (except divide by 0), as long as you do it to **both** sides."

BUT

"If you **multiply** or **divide** an inequality by a **negative** number, you must **reverse** the direction of the inequality symbol."

# SETS

"groups of objects"

$\{1,2,3\}$

order is **not** important

## Types of Sets

(determined by number of elements)

Infinite *Ex.* $\{1,2,3,\ldots\}$

Finite *Ex.* $\{a,b,c\}$

Null *Ex.* $\{\ \}$

## Ways to Show Sets

Roster - an actual list

*Ex.* $\{2,4,6,8,\ldots\}$

Rule - a word description

*Ex.* "all rational numbers (fractions) between 10 and 20, not including 10 and 20"

*Ex.* $\left\{\frac{a}{b} \,\middle|\, a, b \text{ are integers, } b \neq 0, 10 < \frac{a}{b} < 20\right\}$ (set – builder notation)

COURSE NOTE 41

# NAMING SETS

(use uppercase letters)

*Ex.* $E = \{2, 4, 6, 8, ...\}$

---

## Naming Elements of Sets

(use lowercase letters)

*Ex.* "n is in E"

---

## Operations with Sets

Union ($\cup$) – combine the sets

*Ex.*

$A = \{\text{Al, Beth, Carol, Denny}\}$

$B = \{\text{Emma, Beth, Frank, Carol}\}$

$A \cup B = \{\text{Al, Beth, Carol, Denny, Emma, Frank}\}$

Intersection($\cap$) – elements in common

*Ex.*

$A = \{\text{Al, Beth, Carol, Denny}\}$

$B = \{\text{Emma, Beth, Frank, Carol}\}$

$A \cap B = \{\text{Beth, Carol}\}$

COURSE NOTE 42

# RELATIONS WITH SETS

"is equal to" (=) – equal elements

*Ex.* $\{2,4,6\} = \left\{\frac{2}{1}, 4.0, 5+1\right\}$

"is a subset of" ($\subseteq$) – one is contained in another

*Ex.* $\{3,4\} \subseteq \{2,3,4,5\}$

"is a proper subset of" ($\subset$) – one is only a part of another

*Ex.* $\{3,4\} \subset \{2,3,4,5\}$

"is an improper subset of" ($\subseteq$) – one is equal to another

*Ex.* $\{2,3,4,5\} \subseteq \{2,3,4,5\}$

"is a member of"
"is an element of" ($\in$) – an element is contained in another set

*Ex.* $3 \in \{2,3,4,5\}$

COURSE NOTE 43

# VENN DIAGRAMS

(John Venn)

$$\text{For } P = \{c, d, k\}$$
$$Q = \{f, j, k\}$$

$$\text{Universe} = \{\text{letters of the alphabet}\}$$

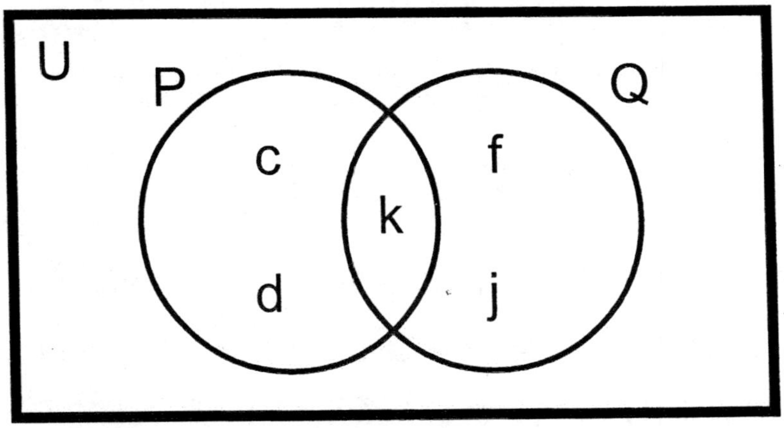


$P \cap Q = \{k\}$ $\qquad$ $P \cup Q = \{c, d, f, j, k\}$

COURSE NOTE 44

# RELATION

"a correspondence between 2 sets"

$$\text{For } A=\{1,2,3,4\} \qquad B=\{5,6,8,9\}$$

$$\text{Domain} \begin{bmatrix} 1\rightarrow 5 \\ 2\rightarrow 6 \\ 3\rightarrow 6 \\ 4\rightarrow 8 \\ 4\rightarrow 9 \end{bmatrix} \text{Range}$$

**"Not a Function"**

---

# Function

"a correspondence (relation) between 2 sets which allows any element of the domain to relate to only 1 element of the range"

$$\text{For } C=\{3,5,7\} \qquad D=\{9,11,13\}$$

$$\begin{bmatrix} 3\rightarrow 9 \\ 5\rightarrow 11 \\ 7\rightarrow 13 \end{bmatrix} \quad n+6$$

**"One-to-One Function"**

COURSE NOTE 45

# OPERATIONS WITH FUNCTIONS

For: $f_1(x) = x+3$ and $f_2(x) = x-1$

$$[f_1 + f_2](x) = (x+3)+(x-1)$$

$$[f_1 - f_2](x) = (x+3)-(x-1)$$

$$[f_1 \cdot f_2](x) = (x+3)(x-1)$$

$$[f_1 \div f_2](x) = \frac{(x+3)}{(x-1)}$$

---

## Relations Between Functions

Equality:

For: $f_3(x) = 2x$ and $f_4(x) = \dfrac{6x}{3}$

$$2x = \frac{6x}{3}$$

So... $f_3(x) = f_4(x)$